How To Make Electric Batteries

by Edward Trevert

with an introduction by Roger Chambers

Self Reliance Books

Get more historic titles on animal and stock breeding, gardening and old fashioned skills by visiting us at:

Introduction

I am pleased to present yet another title on Historical Science.

The work is in the Public Domain and is re-printed here in accordance with Federal Laws.

As with all reprinted books of this age that are intended to perfectly reproduce the original edition, considerable pains and effort had to be undertaken to correct fading and sometimes outright damage to existing proofs of this title. At times, this task is quite monumental, requiring an almost total "rebuilding" of some pages from digital proofs of multiple copies. Despite this, imperfections still sometimes exist in the final proof and may detract from the visual appearance of the text.

I hope you enjoy reading this book as much as I enjoyed making it available to readers again.

Roger Chambers

PREFACE.

THE design of this little volume is to give to the reader the information necessary to make simple yet substantial and practical electric batteries, both closed and open circuit, which can be used for experimental purposes, ringing electric bells, operating telegraph lines, or running small electric motors, incandescent lamps, etc.

The necessary articles and tools required can be easily obtained in any town or city; the expense of them being so small that anybody may afford them.

EDWARD TREVERT.

LYNN, MASS., 1890.

By request of the publishers I have added an appendix to this little book, which I trust will add to its value.

EDWARD TREVERT.

LYNN, MASS., 1903.

Continued demand which has caused the preparation of a third edition of this book has made it seem best to increase its original scope, and to include directions for larger batteries. The earlier sections on familiar types of single cells are retained, as no fundamental changes in battery construction have been made, and the popularity of the book shows the constant need for good directions, such as these have been proven.

CONTENTS.

HOW TO MAKE

ELECTRIC BATTERIES.

ELECTRIC BATTERIES may be classified accord-
ing to their use into open circuit and closed circuit
batteries. An open circuit battery is a battery
which is used when a current is needed for a few
seconds at a time. If the circuit is kept closed too
long the battery will become polarized, that is, hydrogen will collect on the positive plates and prevent the current from passing through the circuit.
If, however, the circuit is opened the battery will
recover itself in time. These batteries are designed for bells, telephones, gas-lighters, etc.

Closed circuit batteries are used for continuous
work, as for electric lighting, electro plating, fire
alarms, etc.

The simplest electric battery made is the Voltaic
Cell. This is made by placing in a glass jar some
water having a little sulphuric acid or any other
oxidizing acid added to it. Then place in it sepa-

rately two clean strips, one of zinc Z, and one of copper C. This cell is capable of supplying a continuous flow of electricity through a wire whose ends are brought into connection with the two strips. When the current flows the zinc strip is observed to waste away; its consumption in fact furnishes the energy required to drive the current through the cell and the connecting wire. The

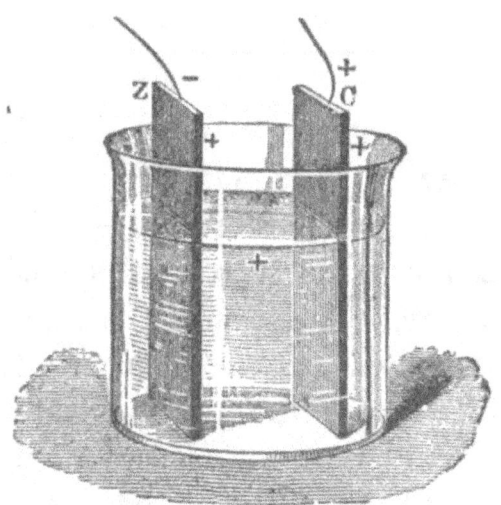

The Voltaic Cell.

cell may, therefore, be regarded as a sort of chemical furnace in which the fuel is zinc. Before the strips are connected by a wire no appreciable difference of potential between the copper and the zinc will be observed by an electrometer; because the electrometer only measures the potential at a point in the air or oxidizing medium outside the zinc or the copper, not the potentials of the metals themselves. The zinc itself is at about 1.86 volts lower potential than the surrounding oxidizing media; while the copper is at only about .81 volts lower, having a less tendency to become oxidized. There is then a latent difference of potential of about 1.05 volts between the copper and the zinc; but this produces no current as long as there is no metallic contact. If the strips are made to touch, or are joined by a pair of metal wires, immediately there is a rush of electricity through the metal from the copper to the zinc, and a small portion of the zinc is at the same time dissolved away; the zinc parting with its latent energy as its atoms combine with the acid. This energy is expended in forcing a discharge of electricity through the acid to the copper strip, and thence through the wire circuit back to the zinc strip. The copper strip, whence the current starts on its journey through the external circuit, is called the positive pole, and the zinc strip is called the negative pole.

This cell however is of little practical use, it being adapted only to experimental purposes, as it will rapidly polarize.

The bubbles of hydrogen gas liberated at the surface of the copper plate stick to it in great numbers, and form a film over its surface; hence the effective amount of surface of the copper plate is very seriously reduced in a short time. When a

simple cell, or battery of such cells, is set to por-
duce a current, it is found that the strength of the
current after a few minutes, or even seconds, falls
off very greatly, and may even be almost stopped.
This immediate falling off in the strength of the
current, which can be observed with any galva
nometer and a pair of zinc and copper plates dipping into acid, is almost entirely due to the film of
hydrogen bubbles sticking to the copper pole. A
battery which is in this condition is said to be
"polarized."

THE LECLANCHE CELL.

This cell is commonly used for what is termed open circuit work—that is to say, for work in which the circuit is open most of the time and the battery sends a current for a short time only between long periods of rest. It has an internal resistance of about 3 ohms, and an E.M.F. of 1.5 volts. It will therefore give a stronger current than the gravity cell, but, as stated above, for only a short time. While not sending a current, it does not deteriorate as does the gravity cell, and it requires almost no attention beyond occasionally filling it up with water, and once in six months or so adding some sal ammoniac. It is not so easily made as a gravity cell but is more convenient for many purposes.

A porous cup of some description must be had to hold the carbon element. It should be about six inches high and two or three in diameter. It must be of unglazed earthen ware in order that the liquid may penetrate it.

Persons who are near a pottery can generally find something suitable there, but when nothing better can be had, a long narrow flower-pot with the hole in the bottom plugged up will do.

The carbon can be sawed from a piece of "gas carbon," which can be obtained from any gas works. It must not be confused with the coke left from making the gas, but is the deposit on the inside of the retorts which is much denser and finer grained than the coke, as well as being purer carbon. A piece must be cut out which will go easily inside the porous cup, and when touching the bottom, project two inches from the top. It should be about twice as wide as it is thick.

The top end of the carbon should be paraffined to prevent the formation of high resistance lead salts between it and its cap. This is done by dipping the end of the carbon into melted paraffine and keeping it there for an hour. The temperature of the paraffine can be kept right by putting the vessel in which the paraffine is melted, into boiling water.

Next drill two quarter-inch holes through the paraffine end, equally distant from each other and the sides of the carbon, and three-quarters of an inch from the end.

We shall next need a mould for the leaden cap. Take a block of wood and make it a quarter of an inch wider and a quarter of an inch thicker than the carbon. Then take a strip of stiff heavy paper and wrap it around the end of the block as shown in Fig 1, the paper projecting an inch and a quarter above the end of the block. First however a copper wire should be tacked to the block in the position shown, the end in the middle sticking up a quarter of an inch, and the wire being pressed close into the corner of the paper where it comes out of the box. Now pour melted lead into this box until it is half full, and then press the paraffined end of the

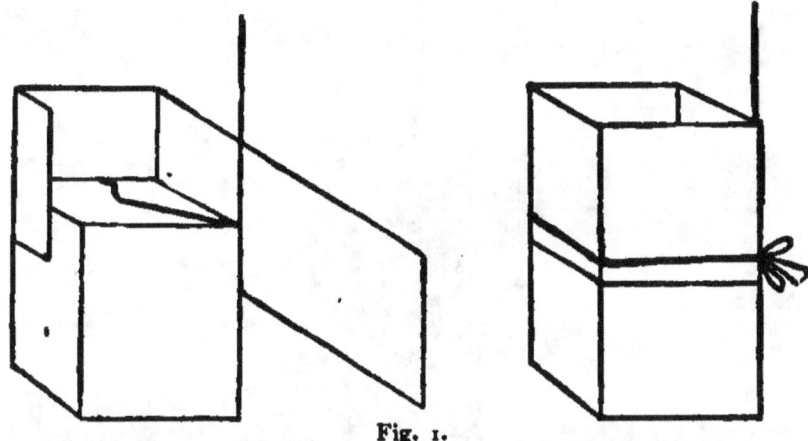

Fig. 1.

carbon into the lead until it rests upon the wire in the bottom of the box and let it stand until cool, when the paper can be removed and the wire attached to the lead straightened out. The lead will run into the holes and hold the cap on firmly, and it should look as in the cut, Fig 2.

If it is not convenient to make a lead cap, a single hole may be drilled through the end of the carbon about three-quarters of an inch from the

end and a round-head wood screw put through it,
and screwed into a wooden block on the other side.
The wire can be placed under the head of the
screw and so clamped to the carbon, see Fig 3.

This however does not make so good a contact
as the first.

Now break up some gas carbon into bits about
the size of a pea, and after sifting them to remove

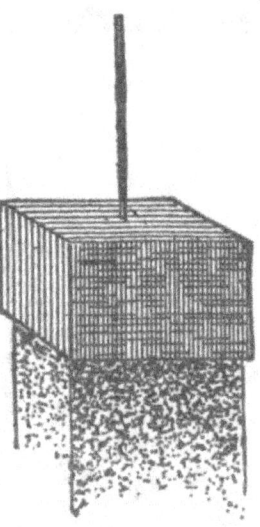

Fig. 2.

the dust, take enough to fill up around the carbon
in the porous cup and mix with it about half as
much peroxide of manganese in the needle form,
which should also be sifted. Pack the mixture
tightly around the carbon in the porous cup.

In the commercial form of this battery the top of
the cup is now sealed over with pitch, a hole being
left for pouring in water to start the battery. A

simpler covering may be made of card board, which is cut out to receive the carbon rod and extends to the edge of the porous cup.

For the zinc element we will need a piece of sheet zinc as wide as the porous cup is long, and long enough so that when rolled up in a cylindrical form it will go around the cup, but not touch it. A wire should be soldered to the top of the zinc.

For our cell we shall need a jar preferably of glass, which shall be large enough to contain the zinc and cup. Put the zinc inside this jar, and inside the zinc cylinder the porous cup.

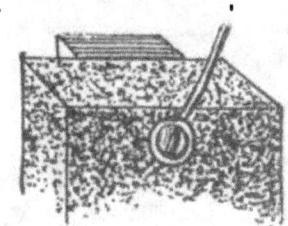

Fig. 3.

The battery complete is shown in Fig 4.

Now put in 4 oz. of sal ammoniac and add water until the jar is three-quarters full, and also pour a little into the porous cup, and leave the battery for three or four hours, when it should be ready to give a current. When the liquid becomes weak, and more sal ammoniac is needed, it will show it by becoming milky in colo1.

Trouble is sometimes experienced with the battery from creeping salts. The paraffine remedy spoken of in connection with the gravity cell can be applied here too.

Another method is to avoid the creeping altogether by using a dilution of sulphuric acid in water (1 part acid to 40 of water) in place of the sal ammoniac solution.

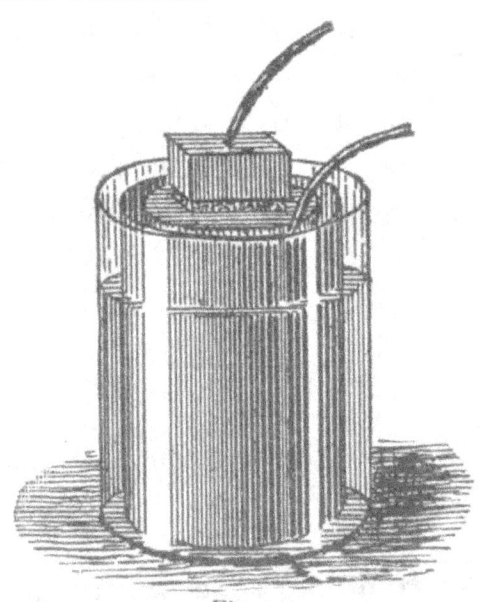

Fig. 4.

A modification of the Leclanche Cell is sometimes used, which is said to have a higher E.M.F., and a lower internal resistance. In place of the broken gas carbon and peroxide of manganese, the carbon rod is packed around with the so called chloride of lime, or bleaching powder, and the liquid instead of being a solution of sal ammoniac, is a solution of common salt.

THE GRAVITY CELL.

For a person with limited facilities at hand, the Gravity Battery will probably present the fewest difficulties in making and will be the most satisfactory for all-around work. Its internal resistance is comparatively high (generally about 8 ohms) and it has an electromotive force of only a volt, but it can be used as few others can for producing a steady current for a long time, or in other words, is what is commonly called a closed circuit battery. The elements used are zinc and copper, and the liquids, solutions of copper sulphate and zinc sulphate.

The battery may be cheaply constructed as follows :—Procure a jar the size you wish your cell to be — it may be of glazed earthenware or glass, preferably the latter, so that the condition of the cell may be seen from time to time. It may be of almost any size but it will be well to have it somewhere near eight inches high and six in diameter. The size of the cell governs the amount of current that may be obtained from it, for while the electromotive force of a cell of any size is the same, pro-

viding the elements are the same, the internal resistance is increased as the cell grows smaller,
which will of course reduce the current. So it
would be well not to make the cell too small, and at
the same time it should not be made very much
larger than the dimensions given, since it would
then become rather unwieldy, and the same results,
viz., the lowering of the resistance can be secured
by using two or more cells and connecting them in
multiple. The jar should have a mouth nearly or
quite as large as its body to facilitate placing the
elements inside it. Should nothing else be available, a large bottle can be made to do by cutting off
the neck and upper part.

To do this, make a file mark on one side of the
bottle where you wish to cut it, and take a heated
iron rod, which has been previously bent for a
short distance to follow the curve of the bottle, and
placing the end of the rod at the file mark and the
curved part against the glass where you wish it to
crack, roll the bottle on a board or table. If it
does not immediately crack, let a drop of water fall
on the file mark and then draw the iron around the
bottle when a crack will follow it.

The jar being provided, next get some sheet
copper for the copper element. It is not necessary
to have this very thick as it is not consumed in the
battery. For the size jar spoken of above, the

copper should be a strip two inches wide, and may be rolled up in the form of spiral, as shown in Fig. 5.

A copper wire should be soldered or rivetted to the copper strip long enough to reach outside of the jar, and it should be insulated to prevent contact with the zinc element. The zinc element

Fig. 5.

should be heavier than the copper as it is consumed while the battery is in action.

To make it, lay out a star-shaped figure on a board, by first drawing a circle on it, half an inch less in diameter than the inside diameter of the jar. Lay off the length of the radius of this circle around the circumference, and you will obtain six points from which to draw lines to the centre of the circle, as in Fig. 6.

With a round chisel gouge out grooves along
these lines, say an inch wide and three-quarters
deep, and you have a mould for the zinc element.
Melt up some zinc in an iron ladle, and after stand-
ing a copper wire up in the centre of the mould as
shown in Fig. 7, pour in the zinc.

When it is cool it should be dipped in acid to
clean it, or if acid is not at hand it should be thoroughly rubbed with sand-paper and then a little
mercury rubbed on it until the whole surface pre-

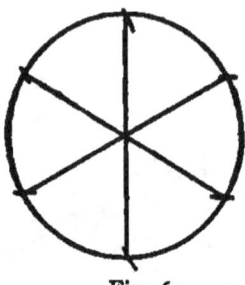

Fig. 6.

sents a smooth and bright appearance. This is
called amalgamation, and while not absolutely es-
sential is advisable in order to prevent what is
called local action, which consumes the zinc without furnishing any useful current.

The wire from the top of the zinc can now be
run through a square stick and bent at right angles
on the other side to prevent its being pulled back
by the weight of the zinc, and the zinc is now
ready for use. In running the wire through the
stick its length should be so adjusted that when

the stick rests upon the top of the jar, the top of the zinc is an inch below the top of the jar.

Fig. 8 shows the elements in place in the jar. An easier way to make the zinc element, is to take sheet zinc such as is used to go under stoves and cut a strip an inch and a half wide. Its length will vary with the size of the jar, and can best be found by trial. Bend it as shown in Fig. 9,

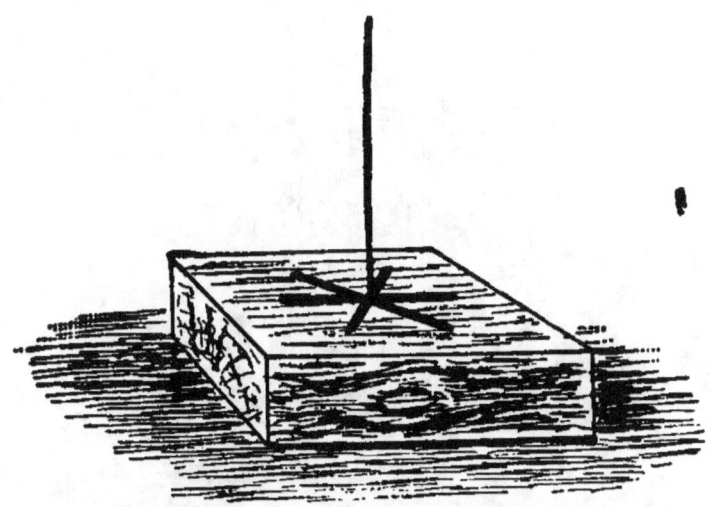

Fig. 7.

leaving the ends long enough to go over the top of a stick and allow the zinc to hang with its top an inch below the top of the jar when the stick is resting on it. The ends of the zinc should be screwed to the stick with round-head wood screws and a wire placed between the head of one of the screws and the zinc.

This style of zinc element will give good results

while it lasts, but that will not be very long if the battery is used constantly. It is, however, easily replaced, and where the making of the other kind is not convenient, will be a fair substitute for it.

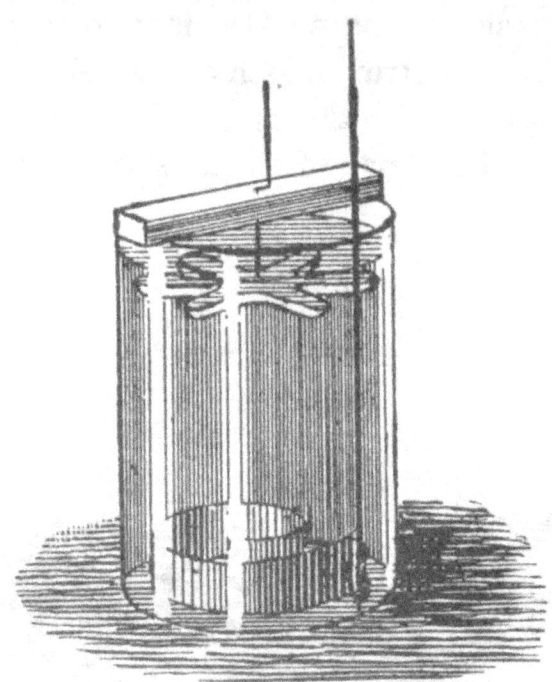

Fig. 8.

The battery is now ready to set up. Place the copper element in the bottom of the jar and around it enough crystals of copper sulphate, or blue vitriol, to half cover it. Then put in the zinc, letting it hang by its stick placed across the top of the jar. Fill up the jar with water until the zinc is covered, and immediately short circuit the battery,—that is, connect the wire from the copper to that from the zinc. Let the battery stand so for six hours when it should be ready for use. A little zinc sulphate added to the water will hasten the action. The copper sulphate on dissolving will give a blue ap-

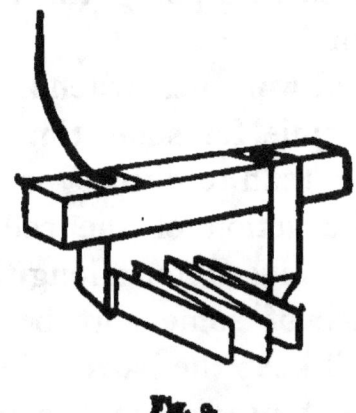

Fig. 9.

pearance to the water in the lower part of the jar. The condition of this blue part is an index to the condition of the battery. In a well kept cell it will rise about half way between the copper and zinc. It should never be allowed to touch the zinc for if it does it will deposit copper on it and this will injure the action of the cell. If the blue line is getting too high it can be made to lower by temporarily short circuiting the battery. If it is too low, that is, so that the copper shows above it, more crystals of copper sulphate must be added and a little of the clear liquid at the top drawn off and water added to take its place.

An annoyance which will probably be experienced is from what is termed "creeping" of the salts. A mass of white crystals will form about the rim of the jar, and will spread unless checked, until they cover the outside of the jar. This may be partly prevented by dipping the rim of the jar in melted paraffine.

The battery will work best when kept constantly at work. If left idle for some time the blue and white liquids will tend to diffuse or mix, which is detrimental to the action of the cell. If the battery is to be left unused for a long time it is best to put a resistance of some sort between the terminals, which will keep the battery in order and at the same time not use enough current to run it down.

A BICHROMATE CELL.

FOLLOWING are given directions for making a good and efficient Bichromate Cell. Take a fruit jar that will hold about a quart, and to this fit a wooden cover made of about one-inch board. Take two carbon plates about six inches long, one and one-half inches wide and one-quarter inch thick, and dip one end of them into melted paraffine wax to keep the salts from creeping. When they are cool, fasten the paraffine ends into the wooden cover. This may be done by cutting two holes about one and one-half inches apart, one for each carbon plate, so that they will fit tightly. The carbons should be connected together by a piece of copper wire inserted between them and the wood. Now bore a hole in the centre of the wooden cover, between the carbon plates, for the zinc. A common Leclanché zinc rod will be good enough for the negative element, but it should be amalgamated. This may be done in the following manner. Take a little sulphuric acid and give the zinc rod a bath in it, then rub a little mercury into the zinc with a tooth-brush, when it will be ready to use. Now screw a binding-post through the wooden cover into the top of one of the carbon plates, and your battery is complete.

HOW TO MAKE A DRY BATTERY.

THE cell is contained in a sheet zinc (all one piece) jar about two and one-half inches diameter and eight inches deep, the metal being about one-sixteenth inch in thickness. A carbon plate five-sixteenths of an inch thick, one inch wide and six inches long, forms the positive electrode. The space between the carbon and zinc is filled with a paste of the following composition:—Oxide of zinc, 1 part, sal-ammoniac, 1 part, water, 2 parts, all by weight. Do not let the carbon touch the zinc, either on the sides or bottom of the cylinder. The unoccupied space at the top may be filled with a mixture of melted beeswax and resin in equal parts. This will seal it and keep it air-tight. The electro-motive force is about 1.3 volts. For convenience a pasteboard box may be made just large enough to set the battery cell in, and cover it to the top of the zinc cylinder.

HOW TO MAKE A CHLORIDE OF SILVER BATTERY.

A CHLORIDE of Silver Battery is very small, and owing to its expense is used for testing purposes and to run small medical batteries.

It is usually about two and one-quarter inches long and one inch in diameter.

The voltage is 1.1, but the internal resistance being 8 ohms, its current output is very meagre. Procure a glass vial or test tube of the above dimensions.

Through a suitable cork stopper press a rod of amalgamated pure zinc, about one-quarter inch diameter and three inches long.

The chloride of silver is to be cast in the form of a cylinder about one-quarter inch diameter and one and one-half inches long, with a silver wire in its centre for connection, and wrapped in a piece of fine parchment paper. The silver wire passes through the cork and is to be soldered to the zinc of the next cell.

A POWERFUL PRIMARY BATTERY.

THE battery for which directions are here given can be made at a low cost, and will give a maximum output of current. The materials to be purchased are glass jars, porous cups, carbons, zincs, burrs, screws, binding-posts, and sheet copper.

All parts of the cell except the carbons can be purchased ready for use. There is a ring of carbons to be fitted into the glass jar as closely as possible. The ring consists of 6 carbon plates screwed to a band of sheet copper $\frac{1}{32}$ inch in thickness and about $\frac{5}{8}$ inch wide. Bend the strip into a six-sided figure, and fasten the carbons (which must first have their tops immersed in melted parafine) securely to it. The fastening may be by screws and nuts, and should be evenly spaced. Bend one end of the copper strip to a vertical position, and fasten it there by one of the screws.

The porous cup will be selected of a size suitable to go inside the ring of carbons. A Daniell, bottle or Fuller zinc should be used, the first having the advantage of a large surface exposed to the action of the fluid. The zinc must be amalgamated, that is, coated with mercury. The best fluid is the chromic acid fluid. Take of chromic acid 18 parts by weight, water 60 parts, sulphuric acid concentrated 9 parts, chlorate of potash 1 part. Add the chromic acid and water, and then pour in a little at a time the

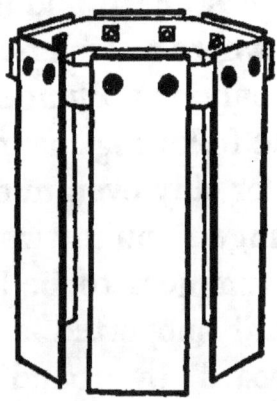

sulphuric acid, allowing it to cool each time. Otherwise the heat will be likely to cause an accident.

When the cell is not in use, take out the zinc, and thus prevent its wasting away.

This is one of the strongest single cells, and with two quarts of solution is capable of considerable work.

PLUNGE BICROMATE BATTERY.

ONE of the simplest, best, and most inexpensive batteries to make where a strong current is wanted for a short time only, is a Plunge Battery. This can be made in the following manner :

Take 4, 6, 10, or any even number of common tumblers, and arrange them in two rows parallel to each other. The tumblers to be held in place by an apertured board supported a short distance above the base-board by round standards. To these fit a board which is split from the standards outward, and provide it with two bolts with wing nuts, by which the board may be clamped at the desired height. Now take as many plates of carbon, about 1 1-2 inches wide, 1-4 inch thick, and 6 inches long, as you have tumblers, and saturate one end of them with a little melted wax or paraffine to keep the salts from creeping. When they are cool fasten the paraffine ends to the opposite edges of the movable board, interposing between the carbon plate and the edge of the board a copper wire. The wooden strips by which the carbons are clamped should be about one-half of an inch thick. Take as many zinc plates as you have carbon plates, taking care to have them the same size as the carbon plates, and amalgamate them. This can be done in the following manner:

Take a little diluted sulphuric acid and give your zinc plates a bath in it, then rub a little mercury into the pores of the zinc with a small tooth brush. Having done this the zincs are ready for use.

Secure the zinc plates to the outside of the wooden strips by which the carbons are clamped.

Fig. 10.

This can be done by ordinary one-half inch screws, passing them through holes in the zinc into the wood. Connect the copper wire of one carbon plate to the zinc of the next, and so on throughout the series. The terminal plates to be connected to two binding posts, which can be made by soldering flat pieces of brass into the heads of round-head wood screws, and screwing them into the wood with the wire under the head. (See Fig. 10.)

Fill the tumblers about two-thirds full of bicromate solution. To make this solution proceed as follows:

To three pints of cold water add five fluid ounces of sulphuric acid. When this becomes cold add six ounces (or as much as the solution will dissolve) of finely pulverized bicromate of potash. Mix well.

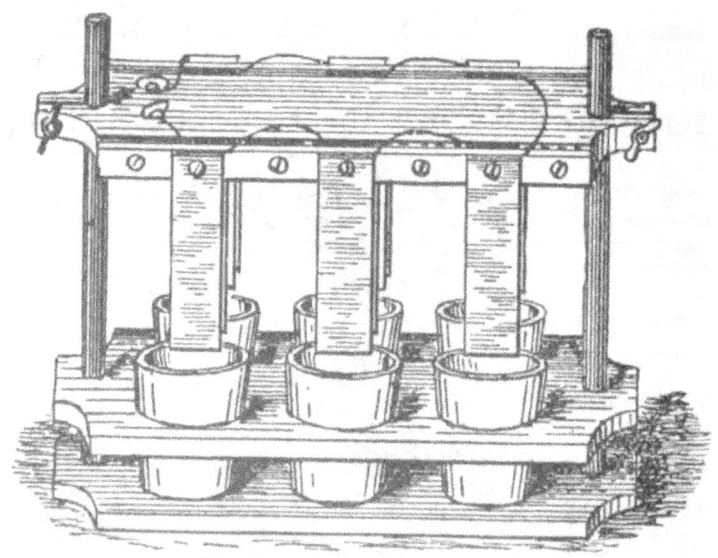

The Plunge Battery Complete.

Always pull your plates out of your solution when your battery is not in use, as the zinc is continually being eaten away when it is in the solution. Each cell will give an electro motive force of about 2 volts.

This battery can be used for running a small incandescent lamp, or motor. A battery of six cells will run a small five or six volt incandescent lamp quite a little while. A much stronger battery can be made by using larger tumblers, or fruit jars and proportionately larger zincs and carbons.

THE STORAGE CELL.

THIS battery has many advantages over the primary battery. For places where a large current is required, such as for running lights or motors, it can be used much more economically than a primary battery, since it has a low internal resistance, and in a well made battery the plates should not deteriorate to such an extent that they have to be thrown away for a long time. It is, however, much more expensive to make, and a good deal of trouble to look after, and when one has not access to a direct current dynamo of some sort, it would be almost out of question to attempt to do any thing with it. However, supposing the reader to be near some electric installation where he can tap a current, we will give him directions for making the battery.

There are two forms of storage battery in use now, that with unpasted and that with pasted plates. We will confine ourselves to the former— for while the latter has many advantages, such as higher E.M.F., greater power for a given weight, etc., the other is more easily made by a person with few tools, and is safer in the hands of a beginner, as it may be short circuited for a little while without damaging it.

A round glass jar (a large fruit jar will do), is the first thing necessary. Of course the remarks made before about the size of the battery and quantity of current to be obtained from it apply here, also. Care must be exercised, however, if the cell is to be large, to get it thick and strong enough to stand the weight of the lead inside without breaking. For the plates we shall need some strips of sheet lead about one-sixteenth or three-thirty-seconds of an inch thick, and about an inch less in breadth than the height of the jar. The length of these strips can best be determined for each case by experiment. Take two of these strips and lay them on a table, as shown in Fig. 11,

Fig. 11.

with paraffined sticks a quarter of an inch thick and three-eighths of an inch wide, in the position shown. Their number will depend upon the size of the cell. They should be placed about three inches apart. The end stick should be tacked to both top and bottom plate, care being taken that the tacks do not go clear through the wood, and

touch the opposite plate, or that two tacks from the
opposite sides do not touch each other.   The other
sticks should be tacked to the top plate only with
two tacks to each stick. Now, beginning at the right
hand, roll up the plates together until you have a
roll which will just go inside the jar, and cut off
what is left. (See Fig. 12.) A stick should be placed

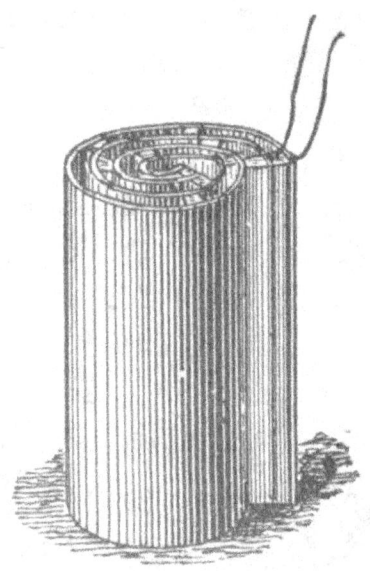

Fig. 12.

between the free ends and tacked to one of the
plates.   Now unroll your plates, and if you are go-
ing to make any more cells of the same size cut
your lead strips by these. Take a coarse file and
laying it upon the lead strip, which in turn is on a
soft board, pound the file with a mallet, moving it
about over the lead until the whole surface has a

roughened appearance.   Then turn it over and repeat the treatment on the other side.   Fix both plates this way, and then roll them up again, being careful that the plates do not touch each other but are kept separate by the sticks, and put them in the jar.   Fill the jar with a mixture of equal parts of water and nitric acid, and let the lead plates stand in it for a day, when they should be removed and thoroughly washed.

This treatment gives the surface of the lead a porous or roughened appearance, and thus increases the surface which may be acted upon by the current.   Now turn out the nitric acid, and place in the bottom of the cell a stick a quarter of an inch square, which has been boiled in paraffine.   This stick is to keep the lead from the bottom of the cell, and thus prevent short circuiting from the fine powder which may fall from the plates.   Solder a wire to the top of each of the lead plates, and put the plates on the stick, and fill the jar with diluted sulphuric acid (1 part acid to 10 of water) which has been mixed four or five hours previously.

In mixing the acid, care must be taken or it may fly up and go over the face or clothes.   The acid should be poured in a fine stream into the water.

Your battery is now ready to charge.   The method of doing this will vary with the circumstances.   If you have a large number of batteries, say 45, and a constant potential dynamo supplying current at 110 volts, the cells may be connected up in series, and placed directly between the mains. When first starting up, or when reversing the cells, it would be well to have a resistance of say 20 ohms in series with the cells which can be cut out when the cells have become formed sufficiently to oppose the current with their normal voltage. This can be seen when one set of plates turn a chocolate color.

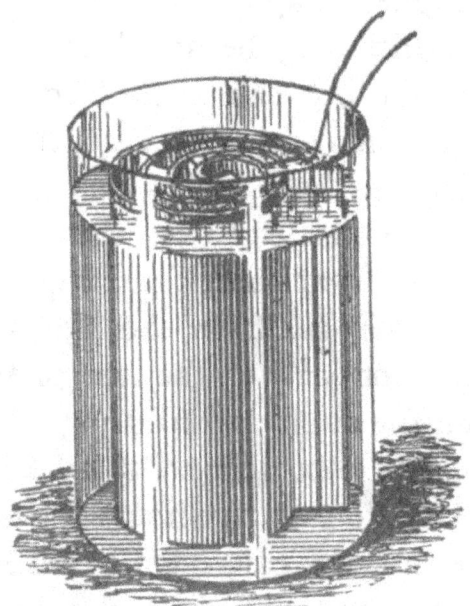

The Storage Cell Complete.

Should you have a fewer number of cells, and the 110 volt circuit at command, insert a resistance in series with the battery, which will cut down the charging current to about 5 amperes for the fruit jar size cell.

When the number of your cells is small, charging from a constant potential circuit of 110 volts becomes very wasteful, on account of the large amount of energy used up on the resistance. Under such circumstances if an arc current is available, it would be well to use that, and in fact it could be used to advantage with a large number as well as a small number of cells.

If it is a current of from 5 to 7 amperes, your cells may be connected in series with each other and the circuit.

If a larger current, say of 10 amperes is used, then the cells should be connected by twos in multiple, and the twos in series in such a way that each cell will receive about 5 amperes.

If you only have one cell and a 10 ampere current, shunt the cell with a resistance of about half an ohm and connect the terminals into the circuit.

We are supposing the most probable cases where a person lives near a lighting station or installation. Any dynamo that can pass a current of from 5 to 10 amperes through the cells may be used, but as there are so many ways in which this can be done, no specific directions can be given that will fit each case. A smaller current may be used, and this is one of the beauties of this cell. It can be charged with a small current for a proportionally longer time, and then give out a large current, for a short time.

Having decided upon the best means of charging our cell we begin by passing a current through it in one direction for from 20 to 30 hours—not necessarily continuously, but preferably so. If the cells begin to boil violently after a short time, the current should be reduced, not that boiling hurts the cell, but it wastes energy. Moderate boiling is not harmful, and is an indication that the cell is working well.

After charging for the specified time, let the cell rest for five or six hours and then discharge it through a resistance that will allow 5 or 10 amperes to pass, and as soon as the cell is discharged begin to charge in the opposite direction, and let it go on for the same length of time as before, after which, let the cell rest again and then discharge. Charge again in the opposite direction and repeat the operations given above until the cell has been discharged four times, when it should be "formed." Now charge again for five or six hours in the same direction that it was last charged and the battery is ready for business.

Each cell should give about 2 volts a little above this at first, and a little under towards the end of the discharge. It should never be allowed to discharge after the E.M.F. has dropped to 1.8 volts per cell. It should never be allowed to stand discharged. After using it for awhile, charge up again, and if the cells are to stand idle for a long while they should be charged every month, to compensate for the losses from leakage.

When a cell is charged it is shown by the liquid boiling briskly. The current passed through it after that is nearly all wasted but a little more may be forced in by stopping the charging current for a while and starting up with the current diminished. After the cell has once been formed the charging should always be done in the same direction.

If the cell has been left discharged for some time, or the acid solution becomes too weak, the positive or chocolate colored plates will "sulphate." or turn a grayish color. This may be remedied by overcharging the battery, which means to continue to pass a current through it after it has commenced to boil. This should be done until the last trace of the sulphate has disappeared.

The battery will lose a great deal of its liquid by evaporation and decomposition, and should be filled up with water as soon as its level gets much below that of the top of the plates.

A SIMPLE AND ECONOMICAL HIGH-VOLTAGE BATTERY.

NUMEROUS attempts, more or less successful, have been made in the past to produce a battery with sufficient power to perform the many practical experiments requiring a voltage of at least fifty.

But in the forms previously used, the great number of cells required, the nuisance of refilling with acids, etc., and the amalgamating of numerous zincs, have made the battery a constant source of work and trouble.

Some have advised the charging of accumulators in multiple from gravity cells and discharging in series, to get the required potential. This plan, on account of the cost of making or buying a secondary battery, has been put out of the reach of many.

It is my aim in this article to describe an outfit which requires care for only 5 gravity cells. The ordinary dry battery answers the purpose of accumulators. This plan, with a suitable switch to change from series to multiple wiring, has worked extremely well with me, and can be quickly and cheaply installed.

The accompanying drawings and photographs will give a very good idea of how such a battery may be set up.

The Switchboard, 2 feet long by 1 and 1½ feet high, has attached to it, about 6 inches from the top, a hard wooden strip 2 feet long, ½ inch thick, and ¾ of an inch wide. On the upper and lower sides of this strip are screwed 20 brass clips ½ inch wide, 1½ inches long, and ½ inch apart. This allows them to protrude ¾ inch to receive a suitably wired stick.

The dry cells, to be used as a secondary battery, are placed on a shelf, 5 cells deep and 8 wide. The cells are arranged in pairs, the 2 cells of each pair being connected in series as shown in the drawing, and the carbons remaining (20) being each connected to one clip on the upper side of the strip on the switchboard, while the zincs are connected to those on the lower side. Care must be taken in attaching the wires to the switchboard that the negative and positive of each pair of cells are connected to clips exactly opposite each other.

Three strips of wood, of the same measurements as the one already attached to the board, are now wired. The first has a strip of brass running nearly full length on both sides (No. 1, Fig. 2) so as to touch all the clips on the upper and lower sides of the switch, thus connecting all the negatives and positives, and when the switch, A, is closed, the dry cells are being charged from the gravity battery.

Strip No. 2 in Fig. 2 has four brass strips on each side, each strip being long enough to touch five of the clips on the switch. The first brass strip is connected with the following one on the opposite side, and so on, leaving the first on the rear side and the last on the side facing, unattached. This rod, when placed in the switch, groups the 20 pairs of cells in sets of five each, connected in multiple. By moving switch L to contacts 1, 2, 3 and 4, these 4 sets may be connected in series, and voltages of 3, 6, 9 and 12 may be obtained.

The last rod, No. 3, Fig. 2, is similar in construction, but made so that each strip of brass touches but one clip. The first brass strip on the side-facing is connected by a small piece of copper wire to the following on the opposite, and so on until only the first on the opposite and the last on the side facing are left unwired. This rod, when placed in the switch, connects all the dry cells in series, and when the switch, A, is open, and the lever, L, is on contact 5, will furnish about 60 volts, provided the dry cells are charged.

In Fig. 1, A is the switch used to connect the primary and secondary, and, of course, this should only be closed when the rod No. 1 (Fig. 2) is in the switchboard. Should this switch be closed when either of the other 2 rods are in the switchboard, the result would be that the current would run back into the gravity cells. The lever, L, should always be on O when the switch A is closed. The binding-posts, E and F, are for use when the primary circuit is required, and the switch, A, is opened.

When fresh dry cells are used, it is not necessary to connect the gravity battery until they become somewhat run down. Care should be taken in wiring the rods and the switchboard that none of the screws are too long and touch.

Such an outfit as this requires only that the gravity battery be refilled when required, and will furnish when fully charged, and when rod No. 3 is in the switchboard, about 60 volts.

It takes some time to charge when thoroughly run down, but for a person only requiring current for a short time each day or so, it is ahead of anything I have tried.

Voltages of 12, 18, 42, and 51 may be had with this rod, by moving switch L over contacts 1, 2, 3 and 4.

CHAS. F. SULLIVAN, *in the Sc. Am. Sup.*, No. 1451.

ARRANGEMENT OF CELLS IN A BATTERY.

THE different methods of connecting the cells of a battery, and their respective uses, is of the greatest importance in battery making, and care in adjustment to the work in hand is well repaid. The three arrangements of cells are *in series*, *in parallel*, and *in multiple*.

IN SERIES. — When a number of cells are connected together so that the positive pole of one cell is joined to the negative of the next — the battery is arranged *in series*, and when so connected it yields the highest electro-motive force of which it is capable. It yields as many times the force of 1 cell as there are cells in the series. That is, if each cell has an electro-motive force of 1 volt, a battery of 12 cells will yield 12 volts.

But the internal resistance of such a battery will, in like matter, be a multiple of that of each of its cells. Thus if the internal resistance of one cell is 5 ohms, the total resistance of the battery will be 60 ohms. But in order to give the best results, the internal resistance of a battery should equal the

Arrangements of Cells in a Battery.

In Series.   One Cell.

In Parallel.

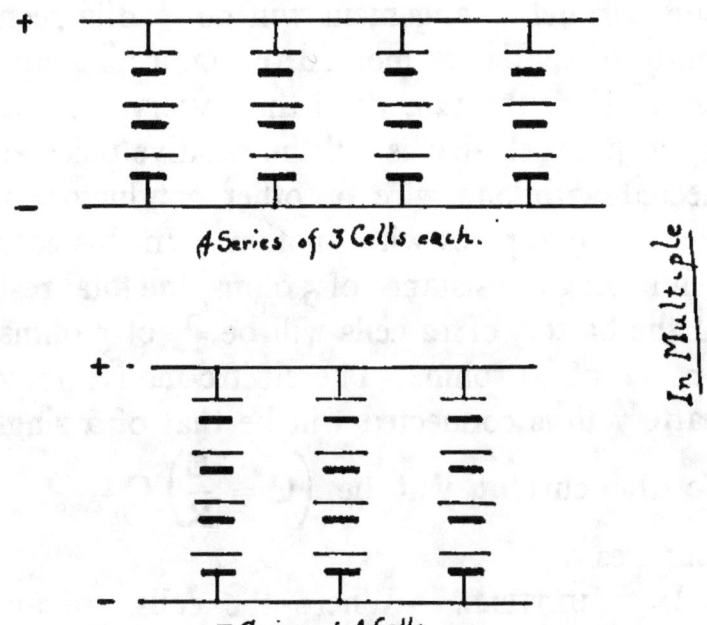

4 Series of 3 Cells each.

3 Series of 4 Cells.

In Multiple.

external resistance of its circuit — so that the series arrangement of such a battery would be of greatest advantage only when that external resistance was 60 ohms also, making a total of 120 ohms.

As the current strength is equal to the electromotive force (in volts) divided by the resistance (in ohms) — we find that $C = \dfrac{E}{R}$ in this case gives us $C = \dfrac{12 \text{ volts}}{120 \text{ ohms}} = \frac{1}{10}$ of an ampere. We have then a current strength of .1 of an ampere, having 12 volts in E. M. F.

IN PARALLEL. — The contrast between the series and parallel arrangement will be readily seen, by a study of the latter method of connecting up a battery. Now the 12 cells of the battery are connected up in parallel, that is, all the positive poles are connected with one wire or other conductor, and all the negative poles with another. In this case, each cell having a resistance of 5 ohms, the total resistance of the battery of 12 cells will be $\frac{1}{12}$ of 5 ohms = $\frac{5}{12}$ or .42 of an ohm. The electro-motive force of a battery thus connected will be that of a single cell. So the current will be $\left(C = \dfrac{E}{R} \right)$ $C = \frac{1}{84} = 1.17$ amperes.

IN MULTIPLE. — Where the cells are arranged three in series, with four such series parallel, as in Fig. 3, the electro-motive force will be 3 volts (this amount remaining the same for any number of series of 3 connected in parallel). The resistance of each series will be 15 ohms, but of the 4 series will be $\frac{1}{4}$ of 15 ohms, that is, 3.75 ohms. Now, making the external resistance of like value, as usual, we find the resistance of the whole circuit 7.5, and by the formula $C = \dfrac{E}{R}$, we get the current strength

$= \dfrac{3}{7.5}$ or .4 of an ampere.

In Fig. 4 the cells are arranged in 3 parallel series of 4 each. The electro-motive force is 4 volts. The resistance of each series is 20 ohms, and of the three in parallel, therefore, $\frac{1}{3}$ of 20 = $6\frac{2}{3}$ ohms. By our formula, then, $\left(C = \dfrac{E}{R}\right)$ we get $C = \dfrac{4}{13.3} = .3$ of an ampere.

It will thus appear that by connecting cells in series, the highest electro-motive force will be secured, but that arrangement is best suited to circuits where the external resistance is high. Cells must be connected parallel for the greatest strength of current, and this method is suitable to circuits of low resistance. By arranging cells in multiple it is possible to obtain a battery suitable to circuits of various resistance, and therefore generally useful.

ZINC FOR BATTERIES.*

As it is considerable trouble for an amateur to cast a zinc plate for his experimental battery, and as an amalgamated one of sheet zinc, without being supported in some way, is too brittle to stand handling, the writer has found that a good zinc plate can be made easily and cheaply by placing an amalgamated piece of sheet zinc between glass plates of the same size, and holding the plates in place by rubber bands.

A short conducting wire should be soldered to the zinc before it is amalgamated. — Bichromate cells for students' work, with zincs one inch by six inches, made in this way, were very satisfactory indeed, and the zincs lasted a long time.

* Thos. R. Baker in Bubier's Popular Electrician.

THE PROPORTION OF CARBON TO ZINC IN A BATTERY.

THE amount of current obtainable from a battery depends largely on its internal resistance, that is the resistance offered by the liquid. The more carbons there are the less the resistance will be.

After determining the amount of zinc that it is cared to use, put in as many carbons as to make four or five times as much carbon as zinc surface.

RULE FOR OBTAINING THE EFFICIENCY OF A BATTERY.

To find the efficiency of a battery, divide the resistance of the external circuit by the resistance of the external circuit plus the internal resistance of the battery, and multiply by 100.

Letting Eff = Efficiency, we have

$$Eff = \frac{R}{R+r} \times 100$$

What is the efficiency of a battery of 50 cells, the external resistance of the circuit being 20 ohms, the internal resistance of each cell being .2 ohms?

Solution : Applying the foregoing rule we have :

$$Eff = \frac{R}{R+r} \times 100 = \frac{20 \times 100}{20 + (50 \times .2)}$$

Equals $\frac{20 \times 100}{30} = 66\frac{2}{3}$ per cent.

HOW SHOULD PRIMARY BAT-TERIES BE CONNECTED WITH STORAGE BATTERIES FOR CHARGING?

IMAGINE a storage cell nearly discharged; it exerts a certain push or electromotive force from its positive pole.

The charging battery must encounter and overcome this electromotive force and implant a charge in the opposite direction. Hence the carbon of the primary (if gravity cells are used the copper) is to be connected to the positive of the storage cells. We do not think it advisable to use anything but glass for jars of gravity batteries. The successful working of these cells is dependent upon keeping the line between the white and blue solutions halfway between zinc and copper. This condition cannot be discernible unless the containing vessel is transparent.

GLOSSARY OF ELECTRICAL TERMS.

ACCUMULATOR.—See battery and condenser.

AMMETER.—An instrument for measuring current strength.

AMPERE.—The unit of current strength. It is the flow of electricity produced by the pressure of one volt on a resistance of one ohm.

ARC.—The stream of hot gasses and particles of carbon visible between the carbons of an arc lamp.

ARMATURE.—That part of a dynamo in which the current is induced. It may be a stationary or moving part, but is generally the latter, and is composed of coils of wire which "cut" the lines of magnetic force produced by the fields. This "cutting" induces a current in the coils.

BATTERY.—One or more cells in which electricity is produced by chemical action. There are two elements of different substances and a liquid in every voltaic battery. A primary battery is one in which the "elements" are placed and used until they are worn out. In a secondary or storage battery or accumulator the "elements" are placed in the cell and first "formed" by the passage of a current of electricity through them. The cell is then said to be charged and may be used to supply electricity. The term battery is also used to designate a collection of Leyden Jars in which static electricity is stored.

BRUSH.—A collection of metal sheets or wires which press against the commutator of a dynamo to collect the electricity, or of a motor to supply it. Carbon brushes are coming into use now, especially in railway work.

B. & S.—Brown & Sharp. The wire gauge used in America.

B. W. G.—Birmingham wire gauge. The English wire gauge.

CELL.—The jar in which the elements and liquid of a battery are placed. The term is used also for the jar and its contents.

C. G. S.—The abbreviation of centimetre, gramme, second, and used to designate the so-called absolute system of measurements.

CIRCUIT.—A system of conductors over which electricity passes.

COIL, CLOSED.—The coils of an armature are said to be closed when the end of one is connected to the beginning of the next at the commutator bar. An open coil armature is one in which each coil is independent of the others and has its own commutator bars.

COMMUTATOR.—That part of a dynamo on which the current from the armature is rectified before passing to the external circuit. The current in a given section of an armature alternates and must be made continuous on leaving it. This is done by the commutator, which consists of a series of insulated metal bars connected to the armature wires, and so placed as to feed into different brushes as the current changes.

CONDENSER.—An apparatus for collecting and holding electricity. It consists of alternate layers of conducting sheets and insulating material, the conductors being very close together, and the adjacent ones being charged with the opposite kinds of electricity. Their proximity enables them to hold a larger amount of electricity than they could if alone. Condensers are sometimes called accumulators.

CONDUCTOR.—A substance which will allow the passage of electricity over it. All substances will do this, but some to so small an extent that they are called insulators.

COULOMB.—The unit of electric quantity. It is the amount of electricity which flows past a given point in one second on a circuit conveying one ampere.

CURRENT.—The flow of electricity in a conductor analagous to the flow of water in a pipe. A continuous current is one that does not change its direction, while an alternating current is one that periodically reverses.

CUT OUT.—An arrangement for interrupting a current or for shunting it around some part of a circuit.

DYNAMO.—A machine driven by power which furnishes electricity.

DYNAMOMETER.—An apparatus for measuring the power given out or consumed by a machine. An electro-dynamometer is an instrument for measuring a current by the mutual action of two coils through which it passes.

ELECTRODE.—A pole of a battery.

E. M. F.—An abbreviation for electro-motive force. This is the pressure which forces the electric current through a conductor.

ELECTRO-MAGNET.—A magnet produced by passing a current through a coil of wire around a soft iron core. The core is magnetized while the current flows, but loses its magnetism when the current stops. This form of magnet may be made much more powerful than a permanent magnet, and is therefore used in place of the latter in dynamos.

FARAD.—The unit of capacity. A condenser that will hold one coulomb at a pressure of one volt has a capacity of one farad.

FILAMENT.—In an incandescent lamp the thread of carbon which becomes luminous when the current is passed through it.

GALVANOMETER.   An instrument for detecting and measuring the electric current by the action of a coil of wire upon a magnetic needle.

INDUCTION.—A current is said to be induced in a conductor when it is caused by the conductor cutting lines of magnetic force. A fluctuating current in a conductor will tend to induce a fluctuating current in another running parallel to it. A static charge of electricity is induced in neighboring bodies by the presence of an electrified body. A magnet "induces" magnetism in neighboring magnetic bodies.

INDUCTION COIL.—An arrangement by which an alternating or fluctuating current in a coil of wire will induce an alternating current in a parallel coil.

INSULATOR.—The opposite of a conductor. A body which will not allow the passage of electricity except in such small quantities as to be negligable.

LINES OF FORCE.—Imaginary lines which radiate from a magnet and show by their direction the path which a free magnetic pole would take if left to itself. Conventionally, the strength of of a magnetic field is indicated by the number of these lines. Their form is shown by the well-known experiment with the magnet and iron filings.

MAGNET.—A body possessing the property of attracting iron, steel and a few other metals.

MAGNETIC FIELD.—The space around a magnet in which its power of attraction is exhibited.

MULTIPLE or MULTIPLE ARC. A method of connecting electric conductors by which a number of sources of electricity feed directly into or a number of receivers of electricity take it directly from the same mains.

NEGATIVE.—A conventional term to indicate the direction of flow of a current, or the state of electrification of a body. The negative or terminal of a dynamo is the one at which electricity enters it from the external circuit, while the negative terminal of a lamp or instrument is that connected towards the negative terminal of a dynamo. It is designated by —

OHM.—The unit of electrical resistance.

OHMS LAW.—States that the current in any circuit is equal to the E. M. F. acting on it divided by its resistance.

PERMANENT MAGNET.—A piece of hardened steel which retains its magnetism after the magnetizing influence is removed.

PARALLEL.—See Multiple.

POLE.—Those parts of a magnet which show the strongest magnetic force. In a bar magnet this is generally a short distance from the ends. The pole of a dynamo or battery is one of its terminals.

POSITIVE.—A conventional term to show the direction of a current. In a dynamo or battery it is the terminal at which the electricity leaves it. It is designated by +.

POTENTIAL.—Power to do work. It is commonly used as synonymous with electro-motive force in speaking of dynamos or batteries.

RESISTANCE.—The opposition offered by a body to the passage of electricity through it.

RHEOSTAT.—An apparatus for throwing a variable resistance into a circuit at will.

SERIES.—Two or more conductors are said to be in series when they are so connected that the same current that passes through one passes through the other.

SHORT CIRCUIT.—An indefinite term used generally in the case of dynamos and batteries for a resistance between the terminals lower than the machine or battery is calculated to stand or run on in practice. With lamps the term is used for a low resistance between the terminals, which deprives it of the most of the current.

SHUNT.—A shunt is a conductor connected around another in such a way that it deprives the first of a part of the current.

SOLENOID.—A hollow coil of wire.

TERMINAL.—The point at which the electricity enters or leaves an electrical apparatus.

VOLT.—The unit of electro-motive force or pressure analogous to the head of water in hydraulics.

VOLTMETER.—An instrument for measuring the voltage or pressure on a circuit.

WATT.—The unit of work. The watts developed in a circuit are equal to the current multiplied by the E. M. F. 746 watts equal one horse power.

WATTMETER.—An instrument for measuring the electrical energy in a circuit.

List of Practical 10c Books

All Books ILLUSTRATED with Working Drawings.

PRICE, TEN CENTS EACH

BUBIER PUBLISHING CO.,
LYNN, MASS., U. S. A.

BUBIER PUBLISHING COMPANY

130 Market Street

LYNN, 𝄪 MASS.

A FIRST BOOK FOR BEGINNERS

in the study of ELECTRICITY and MAGNETISM
in the Popular form of

QUESTIONS and ANSWERS about ELECTRICITY

EDITED BY

E. T. BUBIER

AUTHORS {
T. O'Conner Sloane, A. M., E. M., Ph. D.
Caryl D. Haskins, M. I., E. E.
A. E, Watson, Ph. D., E. E.
Edward Trevert
}

A book by any one of the above authors would find a ready market, for their names bear the same relation to a book on an electrical and scientific subject as does the word sterling on articles of silver. A book by all four is consequently one that every student should have. Then too, this book is arranged in that plain, easy to understand and remember form of "Questions and Answers".

¶ Written in non-technical terms and fully illustrated.

Price, cloth, 50 cents

JUST THE BOOK FOR THE YOUNG STUDENT

12 mo. Cloth 250 Pages 125 Illustrations   Price $1.00, postage prepaid

THE NEW
Experimental Electricity

by EDWARD TREVERT

Sixth edition,   50th thousandth

To the large number of amateurs who are daily springing up in all parts of the country, who desire to perform interesting experiments with electricity and to make the more simple electrical apparatus, this book is dedicated.

You will find much valuable and practical information in this book,—it is devoid of technicalities.

Throughly practical and very interesting.

CONTENTS

Some easy experiments in Electricity and Magnetism.

How to make Electric Batteries.

How to make a Galvanometer.

How to make an Induction Coil.

How to make an Electric Bell.

How to make a Magneto Machine.

How to make a Telegraph Machine Instrument.

How to make an Electric Motor.

How to make a Dynamo.

How to make a Slide Wire Bridge.

How to make a Leyden Jar.

How to make a Wireless Telegraph Apparatus.

How to make a Magnet Winder.

Electric Gas Lighting and Bell Fittings.

Electric Lamps.

Practical Directions for Armature Windings.

Generations of the X-Ray.

Experiments in Metallochromy.

Miscellaneous Experiments.

Something about
X RAYS
for Everybody.

BY EDWARD TREVERT.

A book designed to meet the popular demand for **plain** information regarding this scientific discovery, which **is be-** coming more widely recognized each year, as being **in-** dispensible in treating ailments of the human body, etc., **etc.**

It is intended for the lay reader and is devoid of **tech-** nical language.

Price, cloth 50 cents   **Paper 25 cents**

HOW to MAKE and USE

INDUCTION COILS

BY TREVERT

Revised by B. M. EDMUND, B. S., E. E.

This book is a very valuable addition to the library of the beginner, as it contains, in language devoid technicalities, explanations of some of the fundamental terms and principles of electricity and the practical application of them.

In addition to the theory and description of the parts of an Induction Coil, a detailed description of the construction is given. With the different uses to which the Coil may be put, together with several experiments that may be made, are given general dimensions, so that you may construct the Coil best suited to your individual needs or understand thoroughly the one you have.

Our chapter on Batteries contains descriptions of the approved types of cells and the best method of connecting them to get the desired results.

Fully Illustrated with Detailed Drawings, Diagrams and Halftones

---

Price, 50 cents, Cloth . Paper, 25c

ELECTRIC BELL FITTING and GAS LIGHTING HANDBOOK

BY
EDWARD TREVERT

Giving in plain English complete instructions for installing, connecting and maintaining burglar alarms, annunciators, buzzers, door springs, plain and complicated electric bell and gas lighting systems, the different batteries and their designating features, and the tools and materia. necessary for such wiring.

¶ A Practical Book for Amateurs.

Price, cloth, 50 cents.   Paper 25 cents

THE GAS ENGINE

♣ by P. B. WARWICK ♣

HOW to MAKE and USE IT.

Revised Edition by
L. M. SCHMIDT, Ph. B.

A practical treatment of the ever increasing subject of Gas, Gasoline and Oil Engines

Plain facts about the use, care, fundamental principles and the designating features of the standard gas engines. Technical terms are not used. Just the book for the inexperienced owner of any Gas Engine machine. Fully illustrated and containing detailed drawings and explanations for the building of an improved and practical gas engine and its accessories.

A practical book for practical people

CONTENTS

A Few Historical Points—Principles—Structural and Operative Features. The Design and Operations of different Standard Engines thoroughly covered. Dimensions and detailed drawings for the building of marine engine—table gives dimensions for ¼ to 2½ H.P. and scale drawings are given for a 25 H.P. (actual) engine, which owing to its sensitive govenor is especially adapted to electric light work. How to make a Carburetor, Simple Electric Ignitor, etc., etc.

Price, 75 cents, Cloth

ARITHMETIC
of
MAGNETISM
and
ELECTRICITY

by

JOHN T. MORROW, M. E. and
THORNBURN REID, A. M·, M. E.

This book deals with all those principles of electricity and magnitism which are directly connected with their commercial applications; and to make them clear gives numerical examples based upon them.

A practical book, giving complete rules and tables, and with examples based upon each. Every electrical student must have an Arithmetic, and this is one of the best and most practical. With diagrams.

✚ ✚ ✚

Price, cloth, $1.00

Dynamos and Electric Motors

All About Them
by
EDWARD TREVERT

An illustrated practical book for all who want to construct or understand the principles of Dynamos and Motors.

This book starts in by answering the questions: What is a Dynamo? What is a Motor? Thus it catches the beginners interest at once, and gives him an understanding of the matter in hand. It explains some of the different types of Electric Motors and Dynamos, stating clearly and concisely their various mechanical and electric features.

CONTENTS

Price, Cloth, 50 cents

DIRECT CURRENT DYNAMO DESIGN

Including the detailed calculations upon which the design of a 1 kilowatt Dynamo is based.

by

A. E. WATSON, Ph. D., E. E.

CONTENTS

Detailed Drawings of the Mechanical Construction under :

Armature, Shaft and Pulley.

Base and Field Magnet.

Bearings.

Brush Holders and Yoke.

Commutator.

Spools.

Terminal Board.

Rails.

Detailed Drawings, Diagrams and Calculations of the Electrical Windings, Connections and Assembling with full explanations:

Armature Winding.

Field Winding.

Connections for Shunt Machine.

Connections for Series Machine.

Connections for Compound Machine.

Testing and Using.

Calculations in Detail.

A New Book, describing a dynamo based on the best qualities of earlier designs, and such desiderata of recent types as tooth-drum armature, reasonably slow speed, self-oiling bearings and accessibility of parts.

Price, Cloth, $1.00

✹ PRACTICAL DIRECTIONS ✹
for
ARMATURE
and
FIELD MAGNET WINDING
by
EDWARD TREVERT

While more practicable and particularly adapted for the amateur, it yet contains enough theory to be of value to the more advanced student.

Heretofore standard works on electricity have contained very little information on this highly important subject, for the reason that the art of winding has been mostly theory. This book is, therefore, filling a long felt want of the amateur and student for a reliable practicable work on this subject.

It gives detailed working directions for the winding of Armature and Field Magnets for both Dynamos and Motors. It also gives descriptions of the different apparatus made by several of the leading Electrical Companies in the United States, Useful Tables and other information of value.

FULLY ILLUSTRATED

Price, Cloth, $1.50, post paid

How to Build DYNAMO-ELECTRIC Machinery

Revised Edition by EDWARD TREVERT

350 pages - 200 illustrations - Indexed - Price, cloth $2.50

This book embraces the theory, principles of design and practical directions for the construction of dynamos and motors. It is intended as a practical treatise, and in no way is it to be considered as technical. Some theory, however, is given to help the reader understand principles in a general way. The chapters on commercial types of dynamos and motors are added to show the general construction of large machines. The chapter on management and one on the tables in common use will add to the convenience of the reader.

Full and elaborate directions, together with working drawings and diagrams for the construction of eight or ten practical dynamos are included. The machines have been carefully selected for efficiency and beauty of form. They are easy to build, the design of castings calling for the fewest number of pieces consistent with practical work. Many of these are from the designs of Professor A. E. Watson, which fact speaks for their efficiency, and simplicity. The machines vary greatly in size, from a small but practical toy motor, to a dynamo capable of lighting from 20 to 30 sixteen candle-power lamps.

CONTENTS—Chapter 1, Historical Notes; 2, Principles of Dynamo Machinery; 3, Methods of Field Magnet Winding; 4, Forms of Field Magnets; 5, Armatures; 6, How to Make a Toy Electric Motor; 7, How to Make a Small Dynamo; 8, How to Build a One–Fourth Horse Power Motor or Dynamo; 9, How to Build a Two–Light Dynamo; 10, How to Build a One–Half Horse Power Dynamo or Motor; 11, How to Build a One Horse Power Dynamo or Motor; 12, How to Build a Twenty–Light Dynamo; 13, How to Build a One Thousand watt Alternating Current Dynamo or Motor; 14, Types of Commercial Dynamos—Direct Current; 15, Types of Commercial Dynamos—Alternating Current; 16, Types of Commercial Stationary Motors; 17, Types of Commercial Railway Motors; Appendix A, Management of Dynamos and Motors; B, Useful Tables; C, Some Practical Directions for Armature Winding; D, Field Magnet Winding—Field Formation.

STANDARD BOOKS

Electric Motor Construction for Amateurs. By Lieut. C. D. Parkhurst. Contains 105 pages, illustrated. Just the thing for beginners or anybody wishing to construct his own electrical apparatus. Gives complete directions and working drawings for making an Electric Motor for running sewing machines, small lathes, etc. Also directions and drawings for building an Electric Battery to furnish current for the motor. The amateur interested in motors and batteries will find this book a valuable help. *Cloth $1.00*

How to build Dynamo Electric Machinery. By Edward Trevert. 335 pages, 178 illustrations. This book gives a brief introduction to the principles of construction of dynamos and motors and includes minute directions with working drawings and diagrams for building eight different sizes and styles of machines, from a small toy motor to a 20 light dynamo. It is in every respect a practical book and will be found particularly helpful to the amateur who aspires to construct his own apparatus at home. *Cloth $2.50*

Electricians' Handy-Book of Useful Information. Edited by A. E. Watson, Ph. D., Assistant Professor of Physics, Brown University, Providence, R. I. Compiled by E. T. Bubier. Illustrated. A valuable new book. A compilation from the works of Sylvanus P. Thomson, Kapp, Allsop, Munroe and Jameison, Watson, Bottone, Bonney, Watt, Poole, Trevert, Haskins, Morrow and Reid and others. This book gives a large number of receipts for making Battery Fluids, Battery Pastes, Insulating Materials, Varnishes, Electro Plating Solutions, etc. Tells how to make Electric Batteries, Bells, Telephones, Motors, Dynamos. Induction Coils, Influence, and Static Machines, etc. Gives formulae for winding Dynamos, Motors, Armatures, Field Magnets, Transformers. Contains many Rules, Tables, Data, etc. In fact, it is almost a complete cyclopaedia of Electricity. It is neatly bound and of a convenient size to be carried in the pocket for handy reference. Worth ten times the price asked for it. *Leather, $2.50*

All Books are sent postpaid at the prices listed.

Send for any books not listed, as we can supply any technical book published.

Handy Electrical Dictionary By W. L. Weber M. E. Contains upward of 4,800 words, terms and phrases employed in the electrical profession with their definitions. 2 3-8 in. by 5 1-2 in. and 1-4 in. thick; 224 pages, illustrated, bound in two different styles. Complete, concise, convenient. *Cloth, Red Edges, Indexed, 25 cents. Full Leather, Gold Edges, Indexed. 50 cents.*

Everybody's Hand-book of Electricity. By Edward Trevert. 50 illustrations, 120 pages. Treats of Electric Bells, Electric Lamps, Batteries, Dynamos, Motors, Welding, Etc. Glossary of Electrical Terms and Tables for incandescent Wiring. 50th thousand. *Price: Paper, 25 cents; Cloth, 50 cents.*

How to Build Automobiles. Compiled by E. T. Bubier. Illustrated with working drawings and illustrations. Gives instructions for building both steam and electric automobiles and descriptions of some of the latest machines. *Prices: Paper, 25 cents; Cloth, 50 cents postpaid.*

How to make a Transformer. By L. Milton, Ph. B. Illustrated and with working drawings. *Price, paper, 25 cents*

How to make a Condenser and a Leyden Jar. By T. E. O'Donnell, E.E. Illustrated and with working drawings. Condenser suitable for use with Jump Spark or Induction Coil. *Price, paper, 25 cents*

Telephone Troubles and How to Find Them. By W. H. Hyde. The latest and best book, showing the troubles and the way to remedy them and to keep your telephone in repair. *Price, paper, 25 cents*

A. B. C. of Electricity. By W. H. Meadowcroft. Endorsed by Thos. A. Edison. This excellent primary book is really the first flight of steps in Electricity. *12 mo Cloth, price, 50 cents*

Practical Handbook of Electro-Plating. By Edward Trevert. Gives Practical directions for doing all kinds of electro-plating, using current from either electric batteries or dynamos. Has a chapter on Electro-typing. A plain and sensible introduction to the subject of the electro-position of metals. *Cloth. price, 50 cents*

A Treatise on Electro-Magnetism. By D. E. Connor, E. E. This book is out of the ordinary and is of special interest and value to the student of astronomy, philosophy and the higher laws of nature. *Cloth, price, 50 cents*

DYNAMO AND MOTOR SERIES.

These books give complete directions for building Dynamos and Motors, with working drawings and diagrams.

Cloth 50 cents

No. 1, How to Make a 1-4 H.P. Motor or Dynamo.
No. 2, How to Make a 1-2 H.P. Motor or Dynamo.
No. 3, How to Make a One H.P. Motor or Dynamo.
No. 4, How to Make an Alternating Current Dynamo or Motor.
No. 5, How to Make a 50-Light Dynamo or 4 H.P. Motor.

Electrical Instrument Making. By Bottome
A very complete and practical book for the mechanic. *Cloth 75 cents*

Electric Bells, By Bottone.
A careful study of this book will enable you to wire your house for electric bells and signals of all kinds. *Cloth 75 cents*

Electric Motors By Bottone-Beale.
The study of this book will greatly help the student of electricity. The information contained will save many failures and much time in constructing and understanding electric motors. *Cloth. 75 cents*

How to Make a Dynamo. By Croft.
This book is very popular, and is particularly adapted to the young inventor,—the one who wishes to learn by doing. *Cloth, 75 cents*

TEN CENT SERIES

A series of booklets for the amateur electrician who wishes to make part or all of his electrical apparatus. Each number contains full directions and working drawings for making one piece of apparatus. Descriptions are clear, and vivid, and they contain no technical terms. The drawings give all dimensions, and are worth far more than the price. Start a complete electrical workshop, and make the apparatus yourself. It is far cheaper, and will give you a much better understanding of the subject than any number of books or of machines purchased ready to run.